Colin and the Quest for Quiet Speed

From Takeoff to Supersonic

A Science Story About Flight and Supersonic Speed

by Rudo Kashiri

Curious Minds Explorers Publishing

A Curious Minds Explorers picture book

Ages 8 to 12

Colin and the Quest for Quiet Speed

From Takeoff to Supersonic

A Science Story About Flight and Supersonic Speed

Illustrations by Roshanay

Published by Curious Minds Explorers Publishing

Stories That Launch Curiosity

www.curiousmindsexplorerspublishing.com

ISBN: 979-8-9996203-2-3

Printed in the United States of America

This book is a work of fiction inspired by real science. Names, characters, places, and events are used to encourage curiosity and learning and are presented in a fictionalized format.

This book is dedicated to Ndini, my beloved son, whose fearless spirit inspires me each day. Your willingness to embrace risk and allow your imagination to lead you into remarkable adventures is a constant source of pride. With every opportunity and challenge you undertake, you continue to grow into a true adventurer, an imaginative thinker, and a shining light. This book is dedicated to you with all my love, admiration, and pride.

"Adventure is worthwhile."
— Aesop

This Book Belongs To:

Zoom! Whoosh! Buckle up!

Colin is flying, all the way from Washington, D.C., to Zimbabwe. His heart lifts like a plane at takeoff. And his mind starts buzzing:

- How can a giant airplane stay up in the sky?
- Why does traveling so far take so long?
- Could we go even faster, like a rocket?

Luckily, Colin has Ambuya, his wise grandmother, and a turbo-charged imagination. Together, they'll discover how airplanes fly, how sound travels, and how shape can help a super-fast airplane turn a BOOM into a gentle thump.

Ready, set, science... let's fly!

Colin sat on his bed beside a packed suitcase. He zipped it, unzipped it, then zipped it again, listening to the soft *zzzip* echo through the room.

Baba (Dad) peeked in, smiling. "Excited?"

Colin nodded. "We're really flying to Zimbabwe? Like... across the whole ocean?"

"That's right." Baba sat beside him, and his grin widened. "High enough to see the clouds from above."

Colin's eyes went wide. "How long does it take to get there? Is it really that far?"

Baba held up two fingers, then four. "About 24 hours, from takeoff to landing."

"Whoa... almost a whole day in the sky?" Colin whispered. "How does a giant plane stay up that long? It feels like magic."

Baba chuckled. "It's not magic. It's engineering." He nodded toward the suitcase. "Even packing has a plan."

"We're about to ride on some of the biggest airplanes in the world. There'll be comfy seats, movies, snacks, and drinks. Get ready for an adventure above the clouds."

Colin smiled and sat a little taller, excitement fluttering in his chest. "I have a lot of questions," he said.

"I'm going to ask Ambuya everything!" Colin declared.

Baba laughed and ruffled Colin's hair. "Ask every question you can think of. That's the best way to learn."

"Ambuya!" Colin called, racing into the kitchen. He skidded to a stop. "How do airplanes stay in the sky?"

Ambuya's eyes twinkled. "Ahhh, wonderful question."

She handed Colin a piece of paper. "Let's try a quick science trick. Hold this under your chin, then blow across the top."

Colin held the paper under his chin and took a deep breath. He blew across the top. "It's rising!" he shouted.

Ambuya nodded. "That's lift. A wing makes air move so there's lower pressure above and higher pressure below. The wing also pushes air downward. Together, those help the plane rise."

She held up four fingers.

"Lift pushes up.
Thrust pushes forward.
Weight pulls down.
Drag pulls back."

Ambuya added, "When all four forces balance, an airplane can fly."

Colin's eyes sparkled. He grabbed a sheet of paper. "I'm making a paper airplane right now!"

WEIGHT
DRAG
THRUST
LIFT
Lift
Thrust
Drag
Weight

Colin folded a paper airplane and gave it a gentle toss, watching it swoop, dip, then spin through the air. "So flying is like a balancing game?" Colin said. "That's awesome! It's like everything has to work together."

Ambuya smiled. "It does. The pilot, the engines, and even the shape of the plane work together to keep it steady in the sky. It's like a high-flying dance."

Colin practically bounced off the floor. "But what if a plane went super fast? Could we get to Zimbabwe in no time?"

Ambuya leaned in, lowering her voice. "Sound travels in invisible waves," she said. "Most planes fly slower than those waves, so the sound can spread out ahead. But if a plane goes as fast as sound, or even faster, the plane starts to catch up to its own sound waves. The pressure waves can bunch together and pile up like a crowd at a doorway."

She traced a little wave in the air. "Then those bunched pressure waves can form a shock wave. That's why people on the ground hear one sudden **BOOM**, even after the plane has zoomed past."

"And here's another twist," Ambuya added. "The speed of sound can change depending on how warm the air is... so it isn't the exact same speed everywhere."

She clapped her hands. "**BOOM!** That sound is called a sonic boom."

Colin's eyes went wide. "It's like the sky is clapping!"

Ambuya smiled and nodded. "That's a great way to describe it."

Thrust
Drag
Weight
Thrust
Weight
SONIC BOOM!
BOOM!

Colin zoomed his paper airplane across the table, making a soft *whoooosh* as it swooped past the sugar bowl.

"Ambuya," he asked, "was there ever a real jet that flew super-fast, like in the movies?"

"Oh yes," Ambuya said, smiling. "There was an amazing airplane called the Concorde." She paused, as if she could see it slicing through the sky.

"It flew about twice as fast as many passenger planes, around Mach 2, or twice the speed of sound. You could cross the ocean in just a few hours."

Colin gasped. "Did you ever see one?"

Ambuya's smile turned nostalgic. "I did. It had a long, pointy nose and triangle-shaped wings."

"Engineers in Britain and France built it, and passengers could fly from London to New York in under three and a half hours."

Colin's smile faded into a thoughtful look. "So... why don't we see them anymore?"

Ambuya's voice softened. "The Concorde was incredible, but it was very loud. It was expensive to maintain, and it made powerful sonic booms."

She squeezed his hand. "Many countries didn't allow it to fly over land. The last Concorde flight was in 2003."

"Today, you can visit Concorde jets resting quietly in museums," she added.

Did You Know?

- The Concorde flew up to 60,000 feet, sometimes high enough to see the curve of the Earth.
- It could travel over 1,350 miles per hour (about 2,170 km/h).
- Its nose tilted down during takeoff and landing, so pilots could see the runway.

"Do you want to see something cool?" Ambuya asked as she pulled a few items together.

Colin leaned closer, curiosity bubbling. "What is it?"

Ambuya set a small plate on the table. She stretched plastic wrap across the top like a drum, then sprinkled a few grains of rice on it.

Holding a tuning fork, she gave Colin a playful smile. "Do you want to see sound?"

Colin blinked. "You can see sound?"

Ambuya gently tapped the tuning fork and held it close to the plate.

The rice began to jump, first a tiny wiggle, then a quick hop, then a flurry of bouncy pops, like popcorn that forgot to wait.

"Whoa!" Colin laughed. He leaned in. "The rice is dancing... look at it go!"

"That's because sound is energy in motion," Ambuya explained. "When something vibrates, it sends waves through the air, and those waves can make other things move, too."

She pointed to the plate. "The tuning fork is vibrating, and so is the plastic wrap. That's what makes the rice hop like tiny feet on a trampoline."

Colin tilted his head. "So sound is like invisible ripples?"

Ambuya nodded. "Exactly."

Later, Colin and Ambuya sat close together at the kitchen table, the tablet glowing softly between them.

"Whoa!" Colin said as a sleek jet raced across the screen. "That nose is longer than a crocodile's tail. What is it doing, sniffing out clouds?"

Ambuya tapped the screen. "When a jet flies faster than sound, the sound waves can bunch up and make a big boom."

"But this airplane's long shape helps spread shock waves apart, so people on the ground may hear a gentle thump... instead of a giant **BOOM**."

Colin smiled. "Like tiptoeing through the sky?"

"Exactly." Ambuya tapped the corner of the screen. "This jet is called the X-59."

Colin leaned closer. "So what's the sound barrier? Is it like a speed limit for sound?"

Ambuya tilted her head. "Sort of. People call it the 'sound barrier,' but it isn't a wall. It's a tricky *speed range* around the speed of sound, where airflow changes dramatically."

"When a plane reaches that zone, air waves can bunch up into shock waves, and people on the ground might hear a loud boom."

"The X-59 is designed to shape those waves differently, so the sound comes down softer."

Colin grinned. "So people can sleep right through it!"

Ambuya smiled. "That's the goal."

That's the X-59 a supersonic plane!
X-59
Quiet Supersonic
Less boom.
Less noise.
Faster flight.
X-59

Colin pointed at the jet on the screen, eyes wide. "Ambuya," he asked, "why is the engine on top? Most planes have engines under the wings."

Ambuya smiled and pointed to the picture. "Great question, Colin. When an engine sits on top of the plane, it's a bit like putting a lid on a drum."

She tapped the screen gently. "That placement can reduce some aircraft noise reaching the ground, making it quieter for people below."

Colin frowned, thinking hard. "So... it's the engine's spot that helps make it quiet?"

Ambuya chuckled and shook her head. "The engine placement helps," she said, "but there's more to the story. The biggest reason it's quieter is the shape."

She traced a long, smooth line in the air. "See how the jet looks a bit like a dolphin, long and sleek? That shape helps spread the sound waves out, so they don't pile up in one spot."

She paused, then smiled. "So instead of a loud **BOOM**, people on the ground hear a gentle *thump*."

Colin's face lit up. "So the engine and the shape work together!"

"Exactly!" Ambuya said. "The engine placement helps keep sound off the ground. And the smooth shape softens what people hear below."

X-59 Quiet Supersonic
• Less boom.
• Less noise.
• Faster flight.
X-59
Quiet Supersonic
Less boom.
Less noise.
Faster flight.
X-59

Ambuya pointed to a map on her tablet screen. "Scientists plan community overflight tests over real towns. Then they'll ask people below what they heard: a **BOOM** or a gentle *thump*?"

Colin scooted closer. "How do they test it with people? Do they hand everyone a survey to fill out?"

Ambuya nodded. "Yes. Some people fill out quick surveys. And scientists set up special microphones and sensors to measure the sound."

"Then they compare what people *felt and heard* with what the instruments recorded. That's how they learn if the 'sonic thump' is gentle enough for everyday life."

Colin's eyes lit up. "Will it really be quiet?"

Ambuya nodded. "The goal is for people to hear only a gentle *thump*."

Colin grinned from ear to ear. "Quiet, super-fast flights! Imagine getting to Zimbabwe in no time. No roaring engines, just *ZOOM!* And I'm there."

Ambuya smiled warmly. "And everyone gets to help test it," she said. "By listening... and sharing what they hear. Science works best when people work together."

Did You Know?

The X-59 is designed to fly about Mach 1.4 (about 925 miles per hour).

That night, Colin drifted into a dazzling dream. He was the pilot of the X-59, soaring above the clouds and racing over sparkling oceans and glowing cities.

On the cockpit screen, four glowing arrows stayed perfectly balanced: lift, thrust, drag, and weight.

The world below blurred so fast, even the sun couldn't keep up. "Washington, D.C., to Zimbabwe in just a few hours!" he cheered.

Below him, he imagined only a soft *whump*, like a drum far away.

Colin whooped as he flew through the sky, feeling unstoppable. "I'm flying faster than ever, quiet as a whisper, swift as a rocket!"

"Time to wake up, Colin," Ambuya called.

Colin blinked awake, his dream still sparkling in his mind. He could almost hear that soft *whump* again.

He hopped out of bed and got ready for the flight. He grabbed his backpack and hurried out.

At the door, Colin beamed. "I'm ready, Baba!"

Baba laughed. "That's my explorer."

Colin hugged Ambuya tight. "Bye, Ambuya."

Ambuya kissed his forehead. "Go well, Colin. And keep asking questions."

NASA
X-59

Later, Colin and Baba set off for the airport, traffic humming around them.

In the busy crowd, Baba leaned down with a sparkle in his eye and adjusted the strap of his bag.

Nearby, the boarding line buzzed like a beehive.

Colin hugged his backpack tight, barely able to stand still.

Suitcases clicked and rattled across the floor. Voices swirled, scanners beeped, and announcements echoed overhead. Far off, jet engines rumbled deep and steady, like thunder full of promise.

Baba pointed up at the bright sign above the gate. "This is it, Colin. Ready for your big adventure?"

Colin slipped his carefully folded paper airplane into his backpack, tucking it in like a secret. His eyes shone. "Maybe one day..." he whispered, "we'll fly all the way to Zimbabwe in a quiet supersonic airplane."

Baba smiled and gave Colin's shoulder a gentle squeeze. "You never know. You're a curious explorer. That's how every great journey begins."

As the line crept forward, Colin glanced around the terminal. Bright screens flashed. Feet hurried by. Wheels rolled and hummed.

Then he looked up at the wide sky beyond the windows.

When it was their turn, Colin boarded the plane and found his seat.

His heart raced, ready for the adventure waiting above the clouds.

Gate A12
Boarding Now

As the airplane lifted off, Colin pressed his face to the window, his heart thumping fast with excitement.

Below, the clouds rolled and drifted like endless waves of white.

Colin turned to Baba. "I'm really glad you're on this flight with me," he said. "I can't wait until we land."

Baba smiled and squeezed his hand. "Me too, buddy. We'll land before you know it, and we'll do it together."

Colin grinned, imagining the quiet X-59 flying right alongside their plane, silent, sleek, and impossibly fast.

He reached into his backpack and touched his folded paper airplane, like a tiny promise.

Somewhere in all that rushing air, four invisible helpers were hard at work: lift, thrust, drag, and weight.

In his mind, Zimbabwe waited on the other side of the sky: wide savannas, amazing wildlife, and misty falls.

But mostly, he looked forward to seeing his grandfather, Baba Mukuru.

Colin closed his eyes for a moment, picturing all the adventures waiting in the sky. In his mind, Ambuya's words echoed softly:

"Just imagine what science can do."

Where would you travel if you could fly super fast and super quiet?

After the long flight, the plane tipped gently and glided down. The wheels kissed the runway with a soft bump, and Colin's stomach did a tiny flip.

Warm air rushed in when the doors opened. It smelled like sunshine, warm pavement, and home.

Then Colin saw him. "Baba Mukuru!" He ran forward and disappeared into his grandfather's strong, familiar embrace.

Outside, the sky was wide and bright. High above, a plane traced a thin white line across the blue.

Colin watched it for a moment, then asked, "Do you think one day people will fly to Zimbabwe in quiet supersonic planes?"

Baba Mukuru looked up and smiled. "Every big idea begins with someone brave enough to ask... *What if?*"

He pointed toward the sky. "That's how engineers imagine the future."

Colin grinned. "Ambuya says asking questions helps you learn."

Baba Mukuru nodded. "She's right. Questions are where science begins."

Baba Mukuru gave Colin's hand a gentle squeeze. "Ready to explore amazing Zimbabwe? I'll take you to some of the most incredible places."

"Waterfalls. Ancient ruins. Wildlife. Amazing science." Colin held up his journal. "I'm going to write down every discovery!"

Baba laughed. "First, we get some sleep, Explorer." He put an arm around Colin's shoulder. "Tomorrow, the adventure begins."

ARRIVALS
WELCOME TO
ZIMBABWE

As Colin walked to the car, he felt like a science detective. Runway lights blinked red, blue, and gold. A windsock fluttered and pointed into the breeze. Fuel trucks rolled past like busy helpers.

Farther out, airplanes lined up to taxi, then rolled toward takeoff. Near the runway, a radar tower stood watch.

Everything moved in patterns. Everything worked together.

Colin slowed and looked around. The airport hummed with motion.

It hummed with engineering. It hummed with science.

Science wasn't only in flying.

It was in landing safely, guiding planes, and moving people and cargo, everywhere he looked.

Even the airport was full of science doing its job.

And Colin already had another question, ready to write it down on tomorrow's page in his journal.

Colin's Explorer Journal

WONDERS OF ZIMBABWE

1 VICTORIA FALLS

The water thunders down like the Earth is singing!. The mist feels like rain on my face.

Explorer Note:
The spray can be seen from far away!

2 GREAT ZIMBABWE

These stone walls were built long ago without cement! How did the builders make them so strong?

3 WILDLIFE SAFARI

I saw elephants, giraffes, zebras and so many birds! Every animal has what it needs to survive.

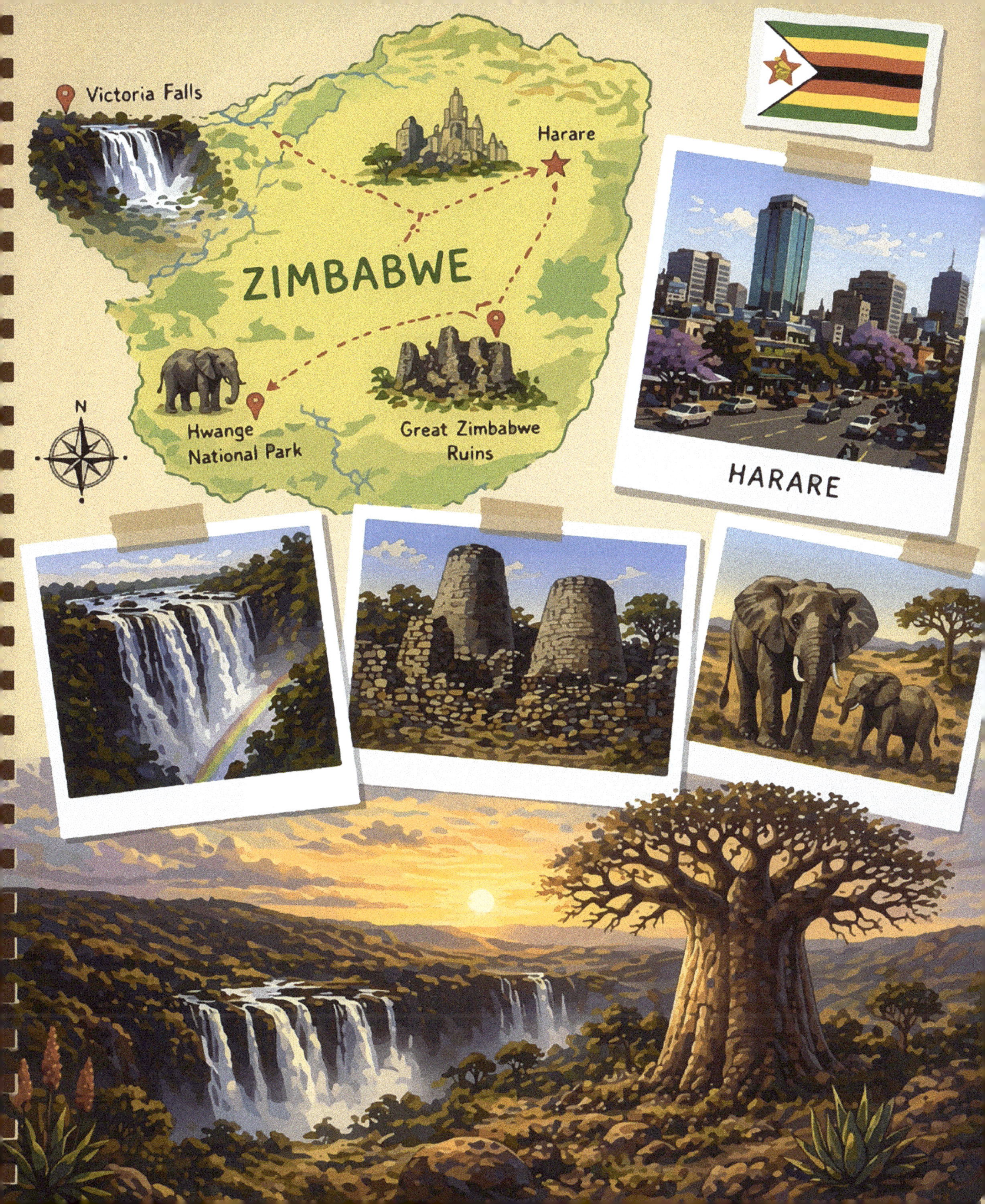
Victoria Falls
Harare
ZIMBABWE
Hwange
National Park
Great Zimbabwe
Ruins
N
HARARE

Ambuya's Corner:
Flying Curiosity Across the World

Every journey teaches us how the world works.

On Colin's flight from Washington, D.C., to Zimbabwe, four forces of flight interact to keep the airplanes flying safely: lift, thrust, drag, and weight. And today, scientists are shaping new aircraft to fly faster than sound while making the boom a gentle thump.

Science connects us and curiosity moves us forward.

Curiosity is your passport. When you ask questions, you can travel farther than you ever imagined.

Four Forces of Flight

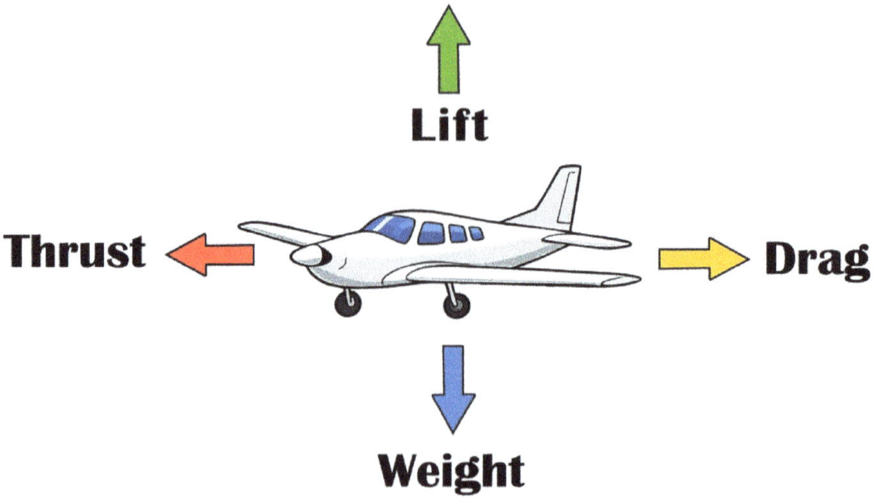

A Note from the Author

Curiosity is the first step toward discovery.

When we ask questions about how airplanes fly or how sound moves, we begin to understand the amazing world around us. Engineers, scientists, and dreamers work together. They ask *what if?* and *why not?*, and they turn imagination into real-world possibilities.

In my work with NASA and as an educator, I've seen how one good question can spark an experiment, a new idea, or even a future career. This book is my invitation to you: stay curious, keep building, and never be afraid to wonder.

I hope Colin's journey inspires you to explore boldly, think deeply, and keep dreaming about what the future of flight and science might become.

Rudo Kashiri

Curious Minds Explorers Publishing

Engineering Design Challenge

DESIGN • TEST • IMPROVE

The Mission: Build a paper airplane, then make one small change at a time to see what happens. Each test helps you discover how the four forces of flight work together: lift, thrust, drag, and weight.

You'll need: 1 sheet of paper + a little open space to fly

Let's fly!

Build + Test

1. Fold your paper airplane.
2. Give it a gentle throw from shoulder height.
3. Watch closely. Does it glide, dip, spin, or stall?
4. Change just *one* thing (wing size, nose folds, or elevator flaps).
5. Fly it again and compare. What changed?

Quick question: Why do you think your change worked (or didn't)?

Now get curious

- Which design flew the farthest?
- Which design flew the straightest?

Engineering tip: Name each version (Plane A, Plane B, Plane C).

www.ingramcontent.com/pod-product-compliance
Lightning Source LLC
LaVergne TN
LVHW070523061126
851857LV00023B/2134